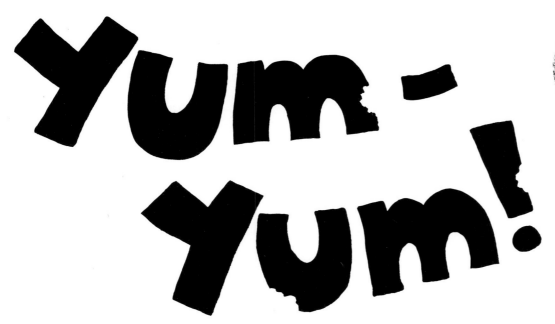

Yum-Yum!

D0183026

MICK MANNING AND BRITA GRANSTRÖM

W
FRANKLIN WATTS
LONDON • NEW YORK • SYDNEY

One night,
out of the soil a tiny shoot
pushes up its head...

Who'd eat a tiny shoot?

The shoot is the first part of a plant to grow above the surface.

Lots of animals would — but caterpillar
gets there first. Yum-yum!

Caterpillar chews up tiny shoot and then crawls across the meadow.
Who'd eat a caterpillar?

Different caterpillars eat different sorts of plants.

Lots of animals would —
but cricket gets there first.
Yum-yum!

Cricket chomps caterpillar,
then hops off through
the grass.
Who'd eat a cricket?

Crickets call in the evening by rubbing their wings together.

Lots of animals would —
but spider gets there first.
Yum-yum!

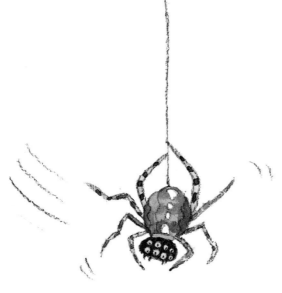

Spider gobbles up cricket and
then dangles by a thread.
Who'd eat a spider?

Some spiders build webs to trap their prey, others hunt on foot!

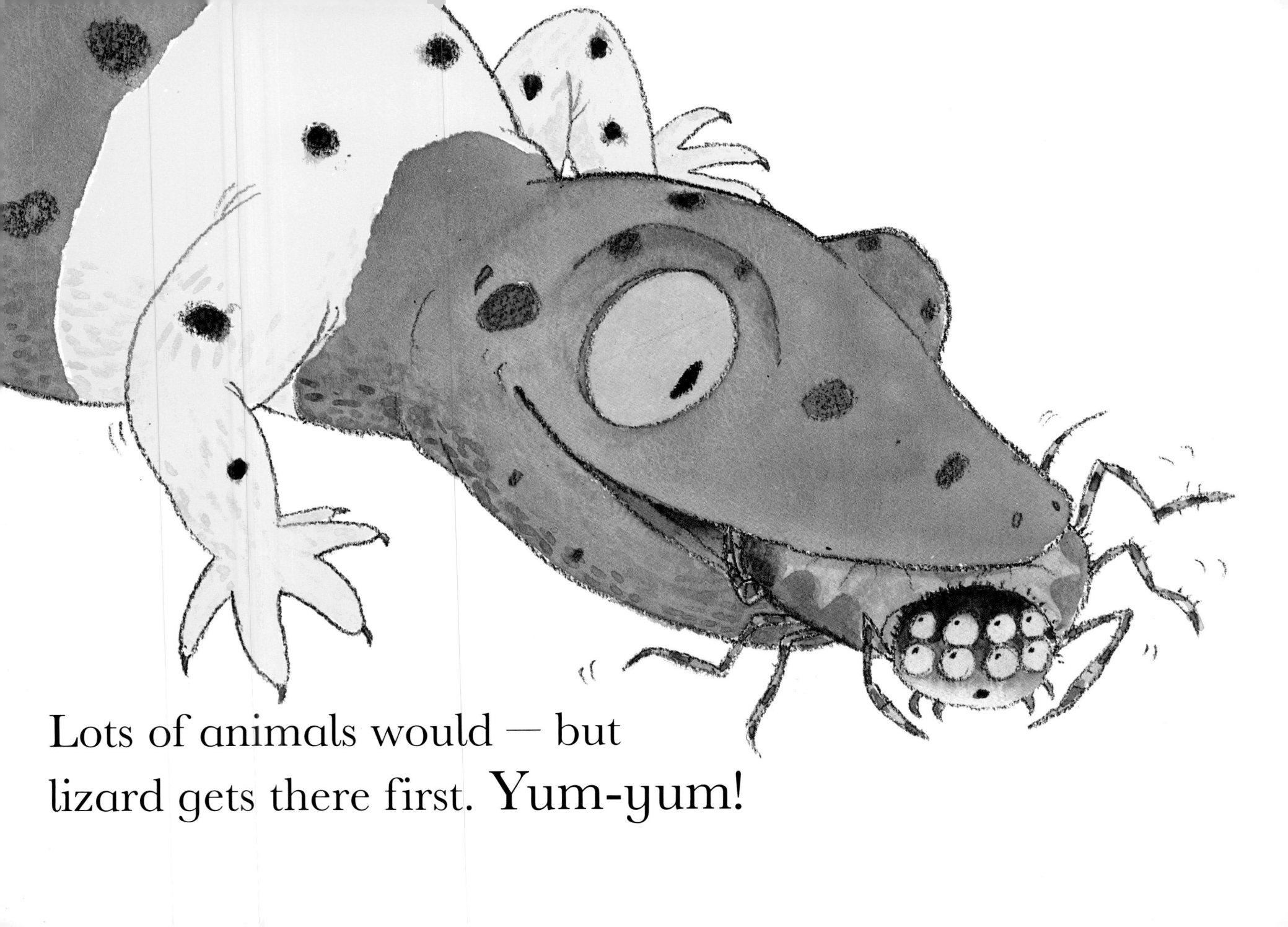

Lots of animals would — but
lizard gets there first. Yum-yum!

Lizard eats up spider, then
sunbathes on a stone.
Who'd eat a lizard?

Lizards live all over the world.
They eat insects and spiders.

Lots of animals would – but little owl gets there first. **Yum-yum!**

12

Little owls often hunt in the daytime, they are very small, about the size of a can of cola!

Owl swallows lizard, then goes to drink water from the lake. Who'd eat a little owl?

Lots of animals would —
but pike gets there first.
Yum-yum!

Pike snaps up little owl, then lazes around on the sunny surface.
Who'd eat a big pike?

Pike will eat other fish, birds and rats!

Lots of animals would — but osprey gets there first. Yum-yum!

Osprey munches pike, then dozes
off on a dead branch.
Who'd eat an osprey?

Ospreys and fish eagles dive into
the water to catch their prey.

Lots of animals would — but fox gets there first.
Yum-yum!

Fox gulps down osprey, but he gets a bone stuck in his throat! Fox coughs and coughs and coughs, then fox dies.
Who'd eat a fox?

Foxes will catch food when they can - but they will also scavenge by roads and in dustbins.

Lots of animals would —
but bluebottle flies and
beetles get there first.
Yum-yum!

Bluebottle flies and beetles lay eggs. Maggots hatch and they eat fox all up, skin and all, until all that's left are foxy specks that sink into the soil! Who'd eat foxy specks in the soil?

Maggots clear up dead animals and rotten food by eating everything up!

Lots of animals would — tiny animals that live in the soil. Yum-yum!

Tiny animals nibble the foxy specks,
making goodness in the soil.
Who'd eat the goodness in the soil?

These small animals are too tiny to see without a microscope!

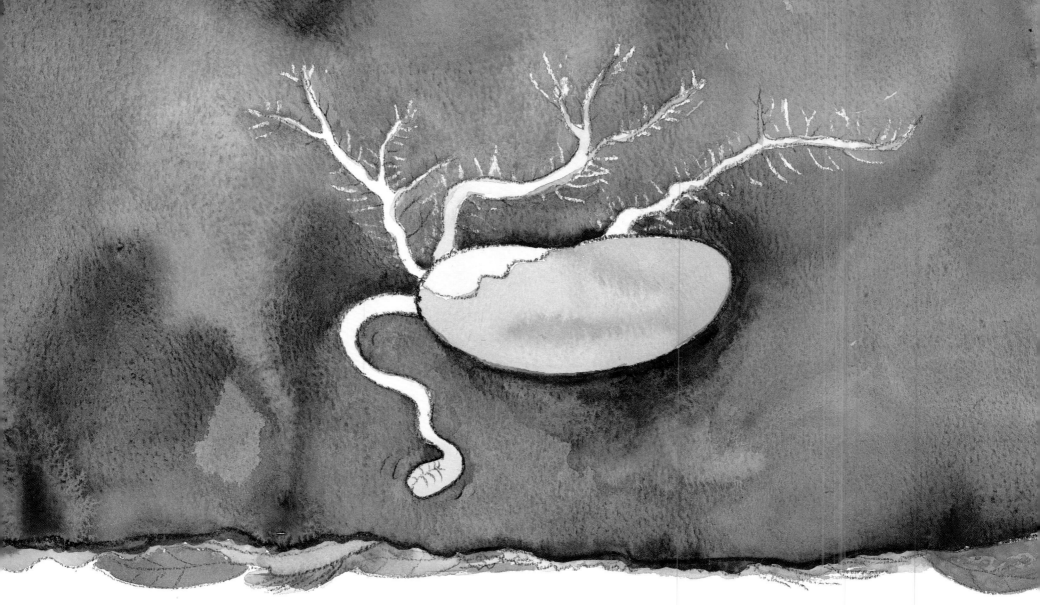

Lots of plants would, but tiny seed gets
there first with its root. Yum-yum!

Tiny seed eats the goodness.
Then one day out of the soil a
tiny shoot pushes up its head.
Who'd eat a tiny shoot ?

It's a race for seeds to grow the fastest and
make use of the goodness in the soil.

Lots of animals
would — but this
time human being gets
there first!

The green shoots grow into wheat.

We make bread from wheat.

YUM-YUM!

29

Helpful words

Bluebottle is a type of fly that lays its eggs on dead animals (pages 20, 21, 29).

Food chain is how a plant and animals are linked together through feeding.

Foxy specks are tiny bits of bone and fur that sink into the ground (page 21).

Goodness in the soil is the natural food that plants need to grow (pages 23, 25).

Maggots are the young of bluebottle flies (pages 20-21).

Magnify is to make something look bigger.

Microscope is a machine for magnifying and looking at tiny living things (page 23).

Predator is an animal that hunts other animals.

Prey is an animal who is hunted for food by another animal (page 9).

Shoot is the first leafy stage of a plant (page 3, 28, 29).

Soil, also called earth, is made up of tiny specks of stones, plants, leaves and dead animals (page 22, 23).

Tiny living things this includes tiny animals such as springtails and mites, and microbes which are sometimes called microorganisms. They are so small you need a powerful microscope to see them (page 22, 29).

For Sue, Steve, Tom,
Emily and Alex.

First published in 1997 by Franklin Watts
This paperback edition published in 1997

Franklin Watts, 96 Leonard Street, London EC2A 4RH
Franklin Watts Australia, 14 Mars Road, Lane Cove NSW 2066

Text and illustrations © 1997 Mick Manning and Brita Granström

Series editor: Paula Borton
Art director: Robert Walster
Consultant: Peter Riley

A CIP catalogue record for this book is available from the British Library

ISBN 0 7496 3130 9 (paperback)
      0 7496 2712 3 (hardback)
Dewey classification 574.5
Printed in Singapore